3D
PRINTING
INDUSTRY

Trends, Markets, and
Players

Timothy J WolF

Table of Contents

Terms Of Use   5

Disclaimer   6

Introduction   7

3D Printing   9

    What it is   13
    Challenges   13
    Creation of demand and availability 14
    Quality and attributes of printed objects 15
    Safety   16
    IP laws   16

Future in 3D Printing: Upcoming Trends   20

    3D Printing Saves Lives   20
    Rapid Product Improvement   21
    Products with Delightful Properties 21
    3D Printing @ Malls   22
    Rising Rights Ownership Concerns   22
    Customization   22
    Novel Machinery Inside Factories   23
    Powder Bed Fusion   24
    Selective Laser Sintering   24
    Melting &   24
    Electron Beam Melting   24
    Inkjet Binding   25
    Photopolymerization   26
    Current Applications   28
    Automotive and Aerospace Sectors   28

3-D Printing in Space   *30*

Medical Application   *30*

World's First 3D Bio-printer   *31*

3-D Printing Penetrates Public Arena   *32*

Open-Source Movement Stimulates Uptake   *32*

Appearance of Online 3D Print Platforms   *33*

The Inexhaustible Age of Mass Personalization   *33*

Unleashing Ground-breaking Possibilities   *34*

3D PRINTING TECHNOLOGY COMPANIES   37

Stratasys, Ltd.   *37*

3D Systems   *38*

Solidscape, Inc.   *38*

Arcam   *38*

Optomec   *38*

Voxeljet   *39*

EnvisionTEC   *39*

ExOne   *39*

Stratasys, Ltd.   *40*

3D Systems   *41*

Solidscape   *44*

Arcam   *45*

Optomec   *46*

Voxeljet   *48*

EnvisionTEC   *48*

ExOne   *50*

Wrapping Up   *52*

REFERENCES   53

TERMS OF USE

DISCLAIMER

The ideas and content contained in this material might not be sufficient for everyone. The author only provides the material as a comprehensive overview of the subject. The author obtained the information from the referenced sources, believed to be reliable, and from own personal experience and research. However, neither implies nor intends any guarantee of accuracy.

All claims made for any product or procedure stated in this book is founded on the author's personal opinion, informed by research. You must do you own careful checking with your own product advisor, referencing other reputable sources, so that your own particular needs and circumstances can be taken into account on any matter that concerns your business or that of others. Research is constantly changing theories and practices in this area. The content of this book is not in any way a statement of professional product or technological advice. Always seek the advice of your product advisors, other business professionals, or technology experts with any questions you may have in this area because they have the necessary knowledge to fit your own needs and circumstances.

The author, publisher and distributors particularly disclaim any liability, loss, or risk taken by individuals who directly or indirectly act on the information contained herein. All readers must accept full responsibility for their use of this material.

INTRODUCTION

Emerging technologies offer answers to critical large-scale challenges and provide a vast ground for sustainable commercial prospects; however, it is essential to find out first and establish which technology trends have maximum potential impact in the near and long-term future. In an increasingly multifaceted, inter- and hyper-connected, risk-aware and see-through world, realizing potential relies heavily on discovering developmental hurdles, allowing receptive and conscientious innovation, and creating collaborations and alliances between governments, business, academics and the public sector.

At some point in the past two decades, advancements in science and technology which are gathering speed spurred a new age of discovery, but at the same time presented huge challenges within existing industries. Betting on emerging technologies, such as the 3D printing technology which has an uncertain market, is a multifaceted proposition that demands profound understanding of value networks and systems. Moreover, public opinion and insight of the potential but inadvertent outcomes of 3D printing can either delay or practically thwart investment.

Existing models for executing technological innovation to prop up economic and social developments are falling short of the challenges afforded by an increasingly intricate and resource-constricted world. As financial systems and business firms become more risk-averse, it is likely for them to overlook opportunities produced by technological progress, and for them to find it increasingly difficult translating innovation into products and tools that address growing global challenges.

If corporations and governments are to guarantee sustainable economic growth, social advancement and a hale and hearty environment, new models for investing in, developing and using 3Dprinting technology are necessary, as well as the creation of a unified research effort that can manage expectations and public perception   issues.

3D PRINTING

Just like the way the personal computer has challenged the prevailing attitude of the conventional computing territory, 3D printing technology has the potential to revolutionize the world of manufacturing. As the accessibility and user-friendliness of 3D printing technology improves while the price plummets, the use of this technology will expand to unprecedented heights. This posits a number of challenges for everyone in the manufacturing and distribution chain simultaneously creating opportunities for new entrants. How we react to these challenges and opportunities will be decisive to its effectivity or collapse.

A Disrupting Technology

3D printing, likewise known as additive manufacturing, is what might best be portrayed as a disruptive technology for the following reasons:

- disruptions of the type the world is now witnessing with additive manufacturing will definitely have a huge impact on business firms with large numbers of workers;
- undersized businesses now have the ability to design and prototype an assortment of products before launching them to market, yet will be able to manufacture these new products on a shoestring. This will result to increased employment for small and medium sized organizations, but not at a scale that will have a major upshot on overall employment;
- huge companies will have to compete with each other with both fewer employees and different types of workers. These employees will be knowledge-based, propelled by innovation, by new concepts and novel methods of interpreting markets;
- markets will become more fluid and a lot more responsive to the changing and, on occasion, volatile needs of consumers;
- the amalgamation of social media and the vastly modified method of commodity production will generally change the balance of manufacturing; and
- the road ahead will see a blend of the bigger conglomerates growing worldwide, with swelling cottage industries that service niche markets gaining strength and momentum.

Taking the U.S. as one example, the U.S. lost six million manufacturing jobs from 2000-2010. These losses will not be recovered just by using traditional manufacturing. The effort then to develop novel types of manufacturing processes that are scaled to leaner economic times is the call of the hour.

Additionally, product lines will have to shift as more information is taken from the prototyping capabilities of 3D printing. To illustrate, take Chrysler that develops and employs numerous car parts which are injection molded. As cases in point:

- Chrysler foresees the company's utilization of additive manufacturing to model over 80% of the products it develops; (Burnett, March 2012)
- SC Johnson puts that same figure at 100%; and,
- Hewlett Packard will employ 3D printing to prototype nearly all the new printers it will generate over the next few years. (Burnett, March 2012)

The primary impact will be in the design of these products; however, as 3D printers scale more effectively to different industries, there will be no reason to delay using them for bigger endeavors. That said; additive manufacturing is only one part of the road ahead as computer-aided design facilitates and promotes the development of new products.

Since the computer is a remarkable starting place to probe into all sorts of possible products where simulation of the real world is conducted, it is easy to test and see if the product has potential. Such practice minimizes risks and eliminates risks. In fact, there are very few products on the market today that have not gone through some insightful assessment using simulation. Assessments range from the mechanical properties of a commodity down to its possible impact within the market place.

Additive manufacturing enables innovative workers to replicate, model, generate and then scale manufacturing to the levels needed for product development and sale. To put 3D printing in further perspective:

- NASA people are contemplating on creating a space station on the moon using in situ 3D printing;
- Dutch architects printing a 3D house; and
- General Electric's decision to use additive manufacturing as an alternative for discrete parts production.

While it is premature to propose significant forecasts in a quantified manner, it is reasonable to say that:

- as design and management groups develop closer working bonds, more disruptive products will be seen and become available in the marketplace than ever before;
- historically, disruptive technologies normally take between 3 - 5 years to have a certifiable impact; and,
- it is practically definite that 3D printing will lead to amplified manufacturing productivity and reduced labor hours per unit of output.

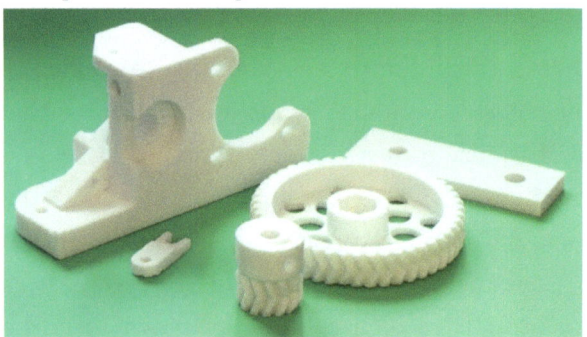

Figure 1 Photo by Andrew Craigie / CC BY

What it is

Basically, 3D printing is merely an expansion, but a very potent and authoritative one, of already existing strategies that utilize intelligent machines to boost human productivity.

It is a process of making a three-dimensional solid object of almost any shape from a digital prototype. The method employs an additive process, where successive stratums of material are laid down in diverse figures/forms, according to a digital design, produced by computer aided design (CAD) or animation modeling software. By "slicing" the blueprint into digital cross-sections, the printer sets down the manufacturing material (liquid, powder, or sheet material) progressively building up layers, until complete. The virtual model and the physical article are nearly indistinguishable

The 3D printing process facilitates quick prototyping, brisk manufacturing and mass customization of products or components, and can significantly trim down new product development time, experimenting and overheads.

The technology has been employed in the production of jewelries, footwear, furniture, weapons, spare parts, and even food, and utilized in the areas of architecture, civil engineering, construction and motor vehicle manufacture, and in the aircraft/aerospace, dental and medical industries.

Challenges

In the next several years, sustained growth and development into industrial, manufacturing, and engineering segments is anticipated. The largest consumers of printers costing more than $5,000 will be service bureaus and manufacturing firms. Considerable growth is also being looked forward to in the hobbyist market via undersized, more economical, and less capable products leveraging online and community distribution. Notwithstanding the apparent benefits like accuracy, the capacity to produce intricate geometries that cannot be generated by other methods, and the comparatively cut-rate price of customized objects, current 3D printing technology does not offer all of the functionality essential to infiltrate the mass market. These limitations include restricted material selection, imperfect quality of printed objects, 'ease-of-use,' and safety of the materials and processes.

Resolving these issues has become major industry focus points. Regrettably, because of the current consolidation in this sector, termination of key patents, and low barriers to entry for new and often inexperienced competitors, "the main inclination is cutting costs for new models of 3D printing equipment based on existing technology. Lower costs are obtained by sacrificing quality, capability, reliability, and safety." (Geller, 2012)

Creation of demand and availability

Development of 3D printers' market demands for the existence of and access to a huge and different population of things to print. In the sudden increase of 2D printing at home, there exist two principal drivers in the creation of things to print. The scale acceptance first was wide- of the PC and software applications that permitted users to create documents without difficulty: word processing, spreadsheets, and presentation graphics. Microsoft Office and related software products permitted consumers to create documents that needed to be printed. The other one was digital photography, which generated an ostensibly endless compilation of things to print.

The availability and ease of use of new CAD products meant for the masses such as Autodesk 123D and SketchUp, as opposed to products from SolidWorks, Pro/Engineer and Catia targeted at the engineering and manufacturing domains) have allowed users to construct models of three-dimensional things to print. Thus, the cosmos of things to print will be greater than before as profit-making organizations and online communities afford models for download, either free or for purchase. The capability then to replicate existing items by scanning or using "photograph to 3D model" software will also boost the available things to print. These can be spare parts, toys, and novelty items, or other tiny articles suitable for home printing.

Quality and attributes of printed objects

Buyers and users nowadays have high expectations for 2D printing quality as exemplified by high resolution color photography. With 3D printing, these expectations will surely be extended. The quality of the printed part must approach the quality of traditional manufacturing methods. Even a naïve user can differentiate 3D printed parts from pieces generated by conventional techniques. Because of the layered manufacturing process, each vertical surface is characterized by a lattice structure artifact. This manufactured article is usually removed by post-processing such as sanding, tumbling parts in drums filled with abrasives, or by solvent vapor smoothing. All these techniques demand a high degree of user interaction, investment in supplementary equipment, and may limit use to industrial environments.

The market will not entirely accept 3D printing until printers prop up numerous materials with physical properties that permit them to print an extensive assortment of items. Unfortunately, most of the existing 3D printers use only one material. Recently commercialized, Objet's PolyJet Matrix technology is capable of generating a mixture of soft and hard materials in the same part.

However, "these parts have low thermal stability, low tear strength, and are prone to rapid aging." (Geller, 2012)

Safety

The single most vital prerequisite for the adoption and implementation of 3D printing by the mass market is safety. The industry has enjoyed considerable flexibility on material safety issues as compared to other industries. This is associated with the low diffusion of the consumer market because the vast majority of equipment is used in industrial environments and models often do not leave the lab.

As the 3D printer journeys toward the home and retail markets, it will die away as an industrial contrivance. Instead, it will be a home appliance. Since these will become home appliances, all build materials (polymers) and specialty agents (chemicals) must be child-friendly and home-safe. Geller (2012) said that plastics must be food-grade and fully-cured (no free radicals) at the completion of the build. Inks and other agents must be non-hazardous and non-reusable as standard household waste. Operation of the device, including maintenance and changing cartridges, must be without mists, continual smell, or outflow.

IP laws

Hall et all (July 2013) wrote that current IP framework seeks to control activity at major facets in the conventional manufacturing and supply chains. As with the introduction of other disruptive technologies that place greater power and control in the hands of consumers (the photocopier, the video recorder and peer-to peer file sharing technology), IP laws strive to sustain with spectacular changes in process or disruption to the supply chain.

Over time, the law was able to respond and take action to these challenges, however, the response was not instantaneous, and as Hall et al (July 2013) stated, in some instances (the case of the Digital Millennium Copyright Act) industry and rights holders lobbied hard for new set of laws.

In February 2013, a team of intellectual property experts within the worldwide Meritas network assembled and organized a panel dialogue, looking at the IP regimes as related to 3D printing in the US, the UK, South Africa and Australia.

3D printing: How to prepare your market and product it threatens

To disregard the threat would improbably generate any considerable business gain. Hall (May 2013) wrote that some brands are embracing 3D manufacturing of simple accessories for their products, as Nokia are doing with 3D printed phone covers. This can be utilized to boost interest in sales for the smart phone itself.

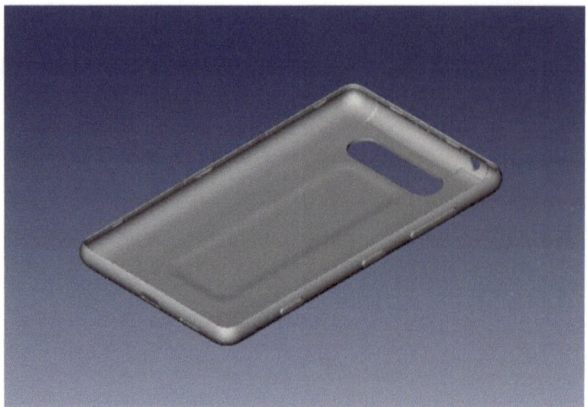

Figure 2 Image: Nokia

Undoubtedly however, that model will be less effective when 3D printers reach the level of complexity and superiority entailed to restructure the multifaceted product itself. Some businesses, however, may fully hold the technology and include, as part of their business model, sanctioned blueprints that permit users to reproduce authorized copies of their products.

Other business firms may concentrate more on the value of the brand and the services provided alongside the original product, persuading consumers "to value the cache and manufacturer support that goes with owning an "original" instead of a 3D printed facsimile." (Hall, May 2013)

Understanding the reality of how encroachment can take place, who may be doing the violation, and what types of products or markets are most likely to be vulnerable will serve users well. Hall et al (July 2013) later concurred that education will assist in putting together practical strategies that will deal with the intellectual property issues better at this point than specific legal and enforcement strategies.

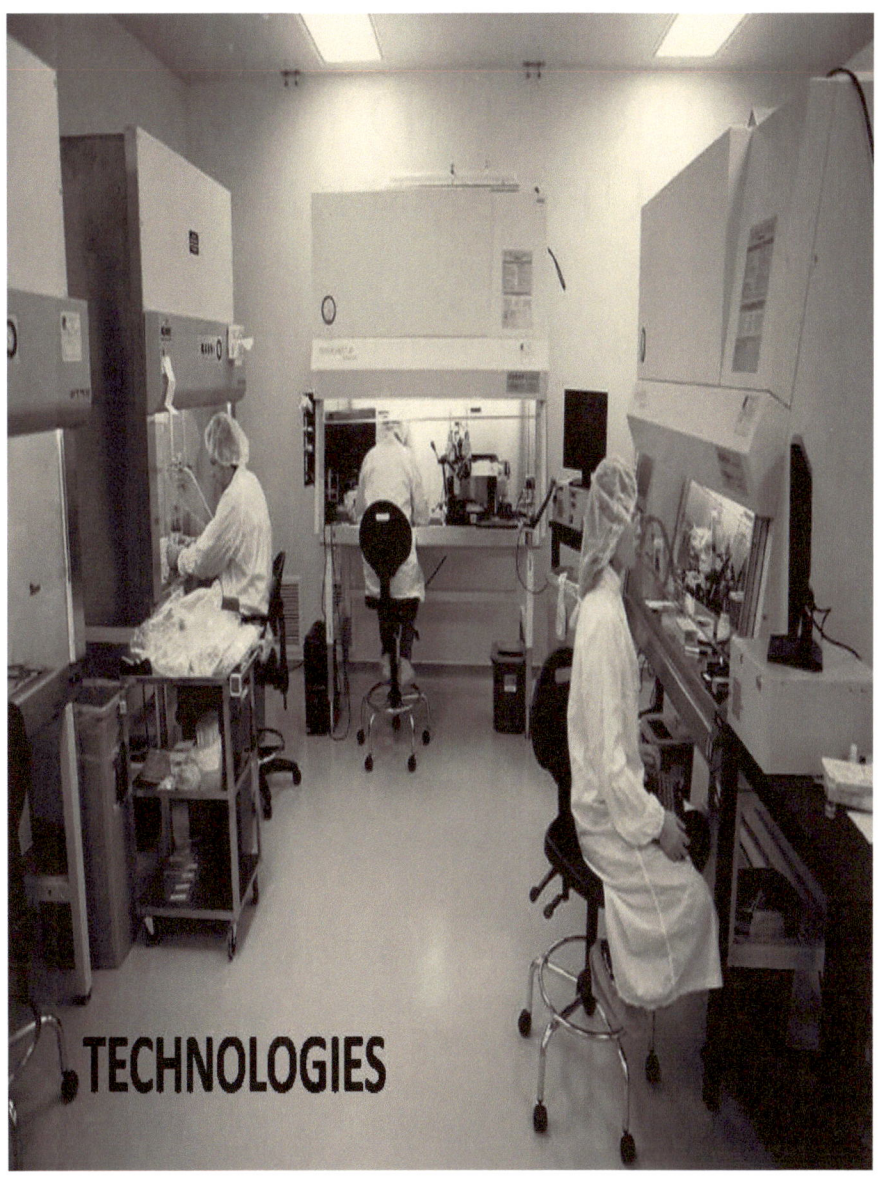

FUTURE IN 3D PRINTING: UPCOMING TRENDS

I f years ago, 3D printing was considered science fiction or just a figment of one's imagination, nowadays, producing objects on demand at comparatively low cost is absolute reality. And such drift is picking up steam in 2013. "With innovative technologies on the rise, here are prevailing trends to watch in 3D printing this year and beyond." (Savitz, Dec 2012)

3D Printing Saves Lives

Savitz (2012) continues that 3D-printed medical implants can advance the quality of life of someone close to you. "As 3D printing facilitates custom-matching of products to an exact body shape" (Savitz, 2012), "it is being employed today for producing more efficient titanium bone implants, better prosthetic limbs and more cost-effective orthodontic implements" (Barcelona 2013). Savitz (2012) also sees that "research and experimentations in printing soft tissue are in progress, and before long can allow printed veins and arteries to be utilized in operations. Current research in medical applications of 3D printing covers Nano-medicine, pharmaceuticals and even printing of organs." Taken to the extreme, there is a strong possibility that 3D printing could enable customized medicines and should reduce, if not get rid of, the shortage in organ donations.

Rapid Product Improvement

With the advent of the 3D printing technology, everything -- from the latest car models to more efficient home appliances – all these will be designed more quickly, bringing innovation to consumers sooner than what they usually expect. Since rapid prototyping using 3D printers lessens the time to convert an idea into a production-ready design, which permits designers to focus more on the function of products. Even though the utilization of 3D printing for rapid prototyping is not new, the fast diminishing expenditure, enhanced design software and ever-increasing range of printable materials signifies that designers will have more access to printers, "permitting them to innovate faster by 3D printing an object early in the design phase, modifying it, re-printing it, and so on" (Savitz 2012). The end-result will be convincingly-enhanced products that are designed faster and done more efficiently.

Products with Delightful Properties

New products that can only be created on 3D printers will be fused in with new materials, nano-scale and printed electronics that will display attributes which will appear thrilling and state-of-the-art compared to today's contrived products. These printed creations will be attractive and viable and will have a unique competitive advantage. Savitz (2012) sees that the nice thing about 3D printing is that it can control material as it is printed, right down to the smallest molecules and atoms. And as today's research is ideal for tomorrow's commercially available printers, the public can look forward to desirable new products with incredible potentials and capacities. Naturally, one basic question would be--what are these products and who will be selling them?

3D Printing @ Malls

Savitz (2012) initially found that these 3D print shops will appear at local markets with first-rate 3D printing services. Since these are designed to service fast-prototyping and other niche capabilities, these shops will branch out to the consumer marketplace. As vendors embark on shipping the design and not the product, the once local 3D print shop will now be positioned where users pick up their customized, locally manufactured commodities, just like the way buyers pick up their printed photos from the local Walmart mall.

Rising Rights Ownership Concerns

As producers and designers begin to wrestle with the potential of their copyrighted designs being reproduced easily on 3D printers, Savitz (2012) believed that "there will be fracas and high-profile test cases over the intellectual property of physical object designs." This will be similar to file-sharing sites that caused so much ruckus within the music industry because they made it so easy to duplicate and share music. The ability to easily make a replica, share, modify and print 3D objects will surely set fire to a new wave of intellectual property issues.

Customization

One can purchase a commodity adapted to one's precise requirements, "which is 3D-printed and delivered to one's front door. Innovative companies will use 3D printing technologies to give themselves a competitive advantage by offering customization at the same price as their competitor's standard products. Initially, this will range from unique items such as custom-made Smartphone cases or ergonomic enhancements to regular devices" (Savitz 2012), however, this will quickly spread out to new markets. Industry leaders will either fiddle with or regulate their sales, distribution and marketing channels in order, as Savitz (2012) writes, "to maximize their capability of providing customization direct to customers. Customization will also play a huge role in healthcare devices like 3D-printed hearing aids and synthetic limbs."

Novel Machinery Inside Factories

Users and markets will witness 3D printing contraptions materializing in factories. In fact, "some niche components are already produced on 3D printers" (Savitz 2012), though this is only on a small scale. Savitz (2012) continues that "numerous manufacturers will initiate 3D printing experimentation for applications outside of prototyping. As the capabilities of 3D printers develop and manufacturers obtain and expand their experience in integrating these into production lines and supply chains," it can be expected that hybrid manufacturing processes will be incorporated with some 3D-printed components which will certainly be further fueled with consumers' desire for products that demand 3D printer-made commodities. With this development, "children will be bringing home 3D printed projects from school as digital literacy (including Web and app development, electronics, collaboration and 3D design) will be backed up by 3D printers in schools. As 3D printing costs continue to fall, more schools will sign on." (Savitz 2012).   Digital literacy will be about things as well as bits.

Powder Bed Fusion
Selective Laser Sintering
Melting &
Electron Beam Melting

"Characteristically referred to as selective laser sintering (SLS), powder bed processing research and development is in progress in many institutions in Europe." (Savitz 2012) This is an AM process where thermal energy carefully combines areas of a powder bed. As with all powder bed based processes, layers of powder are stretched to cover a build platform and facilitate the production of single cross sections (or layers) of the part. This planar, layer-wise character of the process sets aside two dimensional (2D) optical measurements of the bonded layer geometry to be made sporadically for all layers. "Calibration, which includes determination of measurement uncertainty of the optical system, allows quantitative assessments to be made between geometries and measurements." (Cooke, Moynihan; Aug 2011)

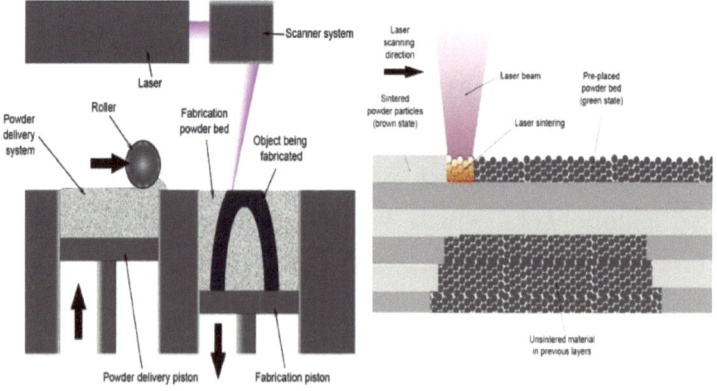

Figure 3 Materialgeeza/Wikimedia Commons

"The selective laser sintering (SLS) was built up, expanded and patented by Dr. Carl Deckard and Dr. Joseph Beaman at the University of Texas in the mid-1980s, with the funding of DARPA. Likewise, a similar process was given exclusive rights without being commercialized by R. F. Housholder in 1979." (Wikipedia, "Selective Laser Sintering"). On the other hand, selective laser melting (SLM) does not use sintering for the blending of powder granules but completely melts the powder using a high-energy laser to create absolutely impenetrable materials in a layer-wise method with analogous mechanical properties to traditional manufactured metals.

"Electron beam melting or EBM is a similar type of additive manufacturing technology for metal parts, for instance, titanium alloys. EBM produces components by melting metal powder layer by layer with an electron beam in a high vacuum. In contrast to metal sintering procedures that operate below melting point, EBM parts are fully dense, void-free, and exceptionally strong." (Wikipedia "3D Printing")

Inkjet Binding

One other technique consists of an inkjet 3D printing system. The printer produces the model one layer at a time by diffusing a layer of powder (plaster or resins) and printing a binder in the cross-section of the part using an inkjet-like process. This is a repetitive process until each layer has been produced. "3D Printing" in Wikipedia states that this technology facilitates the printing of full color prototypes, projections, and elastomer parts. The strength of bonded powder prints can be augmented with wax or thermoset polymer-impregnation.

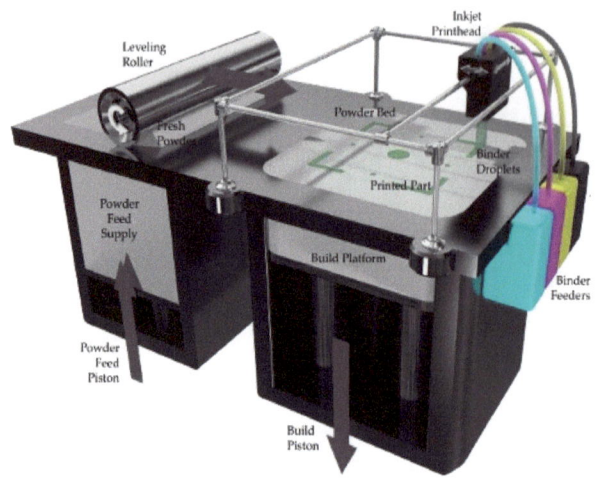

Photo by Virginia Tech

Photopolymerization

Photopolymerization is principally utilized in stereolithography (SLA) to fabricate a rock-solid component from fluids. This method radically redefined preceding endeavors, "from the Photosculpture technique of François Willème in 1860 through the photopolymer process of Mitsubishi`s Matsubara in 1974." (Wikipedia, "3D Printing")

Wikipedia, "3D Printing" provides that in digital light processing (DLP), a container of liquid polymer is exposed to light from a DLP projector under safelight conditions. The liquid polymer that has been exposed to the elements then solidifies.

The build plate travels down in little increments and the liquid polymer is again exposed to light. The procedure repeats until the model has been put together. The liquid polymer is then depleted from the vat, leaving the solid model. The EnvisionTec Ultra is an example of a DLP rapid prototyping system.

Inkjet printer systems, just like the Objet PolyJet system, drench photopolymer materials onto a build tray in ultra-thin layers (between 16 and 30 μm) until the part is completed. Every photopolymer layer is treated with UV light after it is jetted, generating fully treated models that can be held and utilized instantaneously, without post-treatment. The gel-like support material, which is devised to prop up complex geometries, is removed by hand and water jetting. It is also appropriate for elastomers.

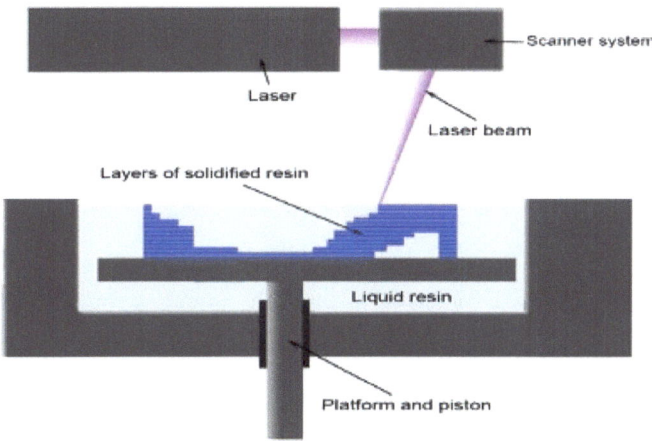

Materialgeeza/Wikimedia Commons

Ultra-small attributes can be made with the 3D microfabrication method used in multiphoton photopolymerization. This method outlines the desired 3D object in a chunk of gel using a focused laser. Due to the non-linear nature of photoexcitation, the gel is treated to a solid only in the places where the laser was focused and the remaining gel is then rinsed away. Feature sizes of under 100 nm are easily fabricated, as well as multifaceted structures with moving and interconnected parts. Nonetheless, another approach employs an artificial resin that is congealed using LEDs.

Current Applications

3-D printing was initially built up and technology expanded for rapid prototyping reasons, making one or two physical samples. It permitted designers to make out and rectify design imperfections fast and inexpensively, thereby stepping up the product development process and reducing business threats.

According to business forecasters CSC, prototyping stays to be the biggest business application of the technology, accounting for some 70% of the 3-D print market. However, advancement in the technology's precision and rapidity, as well as in the quality of materials utilized for printing, have impelled some business sectors to move beyond the employment of 3-D printing in their research and development (R&D) laboratories and integrate it into their manufacturing approach.

This technology is by now extensively used to create jewelry and other custom-made fashion pieces, in dental laboratories to produce crowns, bridges and implants, as well as in the fabrication of hearing aids and prostheses, affording patients a perfect fit. 3-D printing is predominantly appropriate for low-volume, short production runs presenting business firms a more elastic, financially gainful and an immediate option to conventional mass production techniques.

Automotive and Aerospace Sectors

The technology is also being employed to make multifaceted components for the electronics, automotive and aerospace industries. Leading car manufacturers, like GM, Jaguar Land Rover and Audi, have been 3-D printing auto parts for several years now. Major aircraft manufacturers Airbus (part of the European aerospace and defense group, (EADS)) and Boeing are utilizing it to advance the performance of their aircraft and trim down maintenance and fuel overheads. Boeing uses 3-D printing to generate environmental control ducting (ECD) for its 787 aircraft. ECD conventionally demands the production and assemblage of up to 20 various components, but can be 3-D printed in one piece. As further elaborated by one technical officer of a company, additive layer manufacturing is beyond doubt a game-changing technology that has the prospect of revolutionizing manufacturing processes of the 21st century. It can be utilized for an extensive assortment of materials from metals to plastics - including composites - and is faster and more efficient to produce. It employs fewer raw materials and produces components which are lighter, more complex and stronger. Simply put, this is a leaner and greener technology that can be employed by many sectors from aviation through to consumer goods.

Figure 4

Photo by Steve Jurvetson / CC B

3-D-printed aircraft components are 65% lighter but as strong as traditional machined parts, signifying enormous savings and diminished carbon emissions. For every 1kg reduction in weight, airlines save around US$35,000 in fuel costs over an aircraft's life.

To date, aircraft designers already have in their plans the 3-D printing of an entire plane by 2050. To this end, Airbus joined ranks just recently with a South African aviation firm and the Council for Scientific and Industrial Research (CSIR) to probe into the application of titanium-powder-based additive layer manufacturing for making extensive, multifaceted aircraft parts. While costly, titanium is light, sturdy, resilient and long-lasting and perfectly suitable to aircraft manufacture. In conventional manufacturing, it bears machine tools heavily as it solidifies when severed. Such issues are removed in a 3-D print atmosphere.

3-D Printing in Space

NASA engineers have started considering 3-D printing parts, which are structurally more robust and more dependable than traditionally crafted pieces, for its space launch system. For instance, the Mars Rover consists of some 70 3-D-printed custom parts. Similarly, scientists are also probing into the employment of 3-D printers at the International Space Station to make spare parts on the spot. With current trends, what was once the domain of science fiction has now become a definite reality.

Medical Application

The field of medicine is possibly one of the most stimulating areas now using the 3D printing technology. Beyond the use of 3-D printing in producing prosthetics and hearing aids, it is being set out to cure challenging medical conditions, and to press medical research forward, including the field of regenerative medicine. The breakthroughs in these areas are brisk and overwhelming.

One classic example would be the separation of Guatemalan conjoined twins Maria Teresa and Maria de Jesus Quiej-Alvarez in 2002, when doctors at the University of California, Los Angeles' Mattel Children's Hospital employed 3-D-printed models to plan the intricate surgery. By making use of these models, the operation took only 22 hours instead of the expected 97 hours usually required for similar procedures.

Another example was in 2011 when medical doctors at the University Hospital in Ghent, Belgium, successfully performed one of the most complicated facial transplants ever carried out with the extensive use of 3D printing to plan and execute the procedure. Anatomical representations and patient specific guides were 3D-printed for use before and during the procedure.

Then in February 2012, also with the use of a 3-D printer, doctors and engineers at Hasselt University fruitfully carried out the world's first patient-specific prosthetic jaw transplant for an 83-year-old woman experiencing unremitting bone illness. "You can build parts that you can't create using any other technique," says Ruben Wauthle, medical applications engineer at Layerwise, the company that built the implant, in a BBC report. "For example, you can print porous titanium structures which allow bone in-growth and allow a better fixation of the implant, giving it a longer lifetime."

World's First 3D Bio-printer

3D printing technology is also being utilized to grow fresh human tissue. In 2009, Organovo, in partnership with Invetech, created and developed the world's first bio-printer. The MMX takes primary or other human cells and forms them into 3-D tissues for medical research, including drug development and therapeutic applications. And in late 2010, Organovo publicized that it had produced the first bio-printed blood vessels.

3-D Printing Penetrates Public Arena

Outside of these intriguing business applications, 3D printing has begun to filter into the mainstream.

Even though 3D printers are not yet a customary element of home-computing apparatus, the most up-to-date generation of devices, such as Cube® by 3D Systems, the Cubex™ or Makerbot's Replicator™2X - which have market retail prices of between $1000-$3000 - are bringing the likelihood of home manufacturing one step closer to reality.

A study by Wohlers Associates foresees that the sale of additive manufacturing products and services will get to US$3.7 billion by 2015, rising to over US$6.5 billion by 2019.

Open-Source Movement Stimulates Uptake

The uptake and development of 3D printing is also being kindled by the vibrant open source movement. For instance, the RepRap (replicating rapid prototyper) idea, founded by Dr. Adrian Bowyer at the University of Bath, UK, in 2005, has generated an economical 3D printer capable of printing most of its own components. The project's designs, including the machine itself, are released under a free software license, the GNU General Public License.

One major goal of the initiative is to place inexpensive desktop manufacturing systems in the hands of people anywhere in the world, so they can construct multifaceted and intricate commodities themselves with very modest capital outlay. A RepRap kit costs around US$500. As the RepRap printer design is open, anyone can modify or improve, manufacture and sell it. Business analysts have noted that the rate of innovation of the RepRap and its offshoots is gathering speed more rapidly than comparable commercial 3D printers.

Appearance of Online 3D Print Platforms

The number of online 3D print platforms has now escalated, just like the Makerbot's Thingiverse as it makes it possible for individuals to upload and share their designs or download designs for manufacture.

For those without direct access to 3D print technology, an increasing collection of online services are available and can be accessed without much difficulty. For instance, Shapeways and Sculpteo, recommend platforms for people to share their ideas and make them real by giving access to forward-looking 3D software and printers. As of August 2012, Shapeways prided itself of roughly 7,000 shops and over 16,000 members, who had printed over a million products. A collection of software applications, such as Autodesk 123D, is also available for people to design and modify objects on their home computers.

The Inexhaustible Age of Mass Personalization

3D printing is paving the way for an innovative period of mass personalization. In January 2013, Nokia publicized it is making the 3D printable files of its Lumia 820 phone case available to customers, so they can fashion their own designs and print them on any 3D printer. Even as end users are not likely to print what is readily available in the stores, when it comes to making personalized items, widgets or extremely rare pieces, the scope for 3D printing applications is boundless.

Unleashing Ground-breaking Possibilities

In order to reach its full potential as a manufacturing technology, a number of technical hurdles still need to be surmounted, specifically with regards to the cost of materials, quality of output, size restrictions and throughput capacity. As noted by one consultancy firm, 3D printing is offering a stage for cooperation and partnership that is increasing the tempo of innovation and disruption in the material world, just like the way the Internet advanced collaboration, innovation and disruption in the digital world.

Chris Anderson, former Wired magazine editor-in-chief and author of Makers: The New Industrial Revolution explains, "when a technology becomes desktop, it doesn't just get cheaper, smaller, better, more ubiquitous, what happens is it gets used in different ways....It becomes "a vector for ideas which are turned into things, companies, movements and that moment is right now." The purported "democratization" of manufacturing that 3D printing guarantees has enormous potential to let loose the creativity of the masses and encourage economic development and growth.

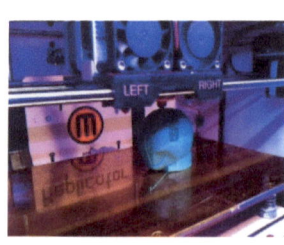

Figure 5 Photo by Nicholas Wang / CC BY-SA

Conventional manufacturing demands high levels of capital investment and large-volume product-runs. By radically trimming down capital expenditure, overheads and business threats, 3D printing can make it more convenient for anyone to be part of the manufacturing process and experiment on their ideas.

While the full repercussions of its extensive acceptance are ambiguous, by making "manufacturing on demand" a rational option, the uptake of 3D printing could convert the manufacturing and business backdrop all over the world. It can lessen the necessity of carrying inventory, slash down warehouse/transport costs, shorten supply chains and extensively cut down the carbon footprint of manufacturing.

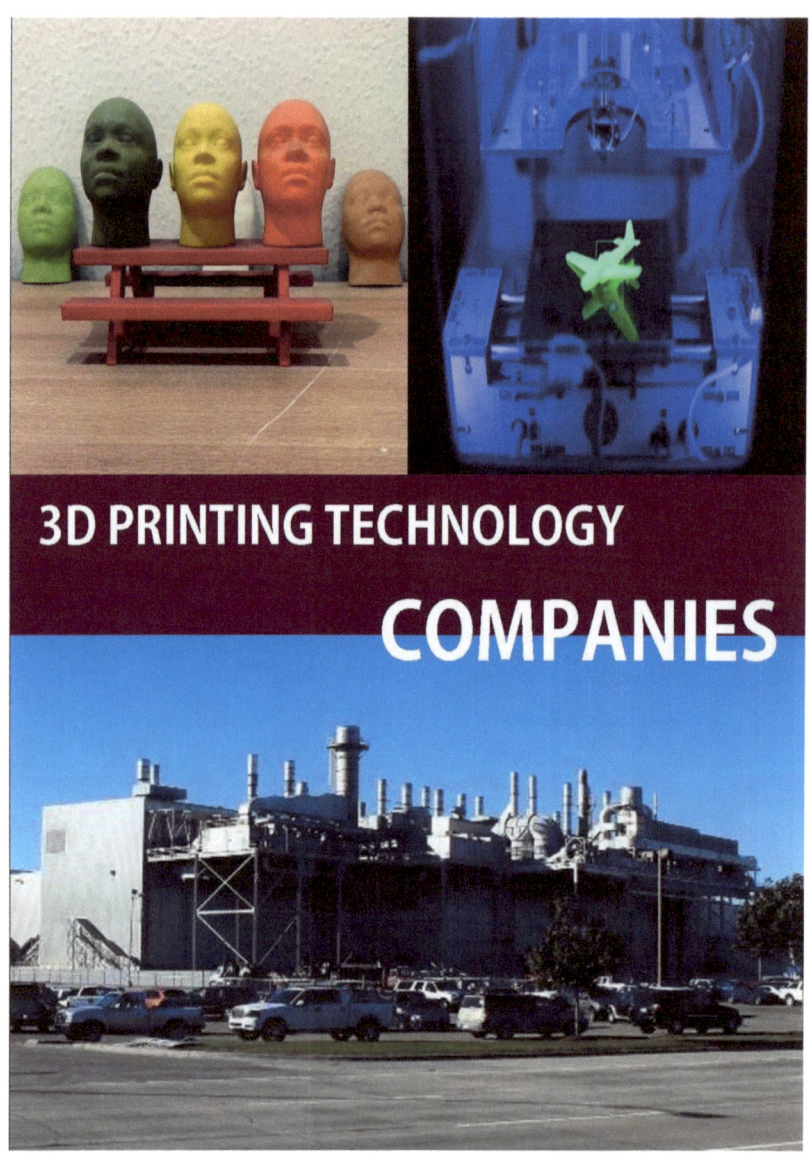

3D PRINTING TECHNOLOGY COMPANIES

The following pages and their sections present individual companies that are noted and known for their contributions to the market and study of 3D technology.

In presenting each company, rather than rewriting or paraphrasing what is already so well stated in each company's own website text, the text used in each company's section is taken as straightforward as possible from that company's website and website pages, and directly related financial advisory sites.

Each company's source website addresses and source site pages are cited below the company items that follow.

~ ~

Stratasys, Ltd.

A producer of 3D printers and 3D production systems for office-based rapid prototyping and direct digital manufacturing solutions.

~ ~

3D Systems

3D Systems Corporation is a holding company incorporated in Delaware and operates through subsidiaries in the United States, Europe and the Asia-Pacific region.

~ ~

Solidscape, Inc.

Solidscape Inc. is a major manufacturer of high-precision 3D printers, materials and software for Direct Manufacturing of solid objects designed through the use of CAD.

~ ~

Arcam

Arcam offers a cost-efficient additive manufacturing solution for production of metal components.

~ ~

Optomec

Optomec, a lucrative and swiftly growing additive manufacturing firm, spearheads the next generation in electronics, energy, life science and aerospace/defense manufacturing.

~ ~

Voxeljet

Voxeljet, another leading manufacturer of industrial 3D printing systems, operates what it believes is one of Europe´s largest service centers for the on-demand production of moulds and models for metal casting.

~ ~

EnvisionTEC

In 2002, EnvisionTEC began creating and generating Perfactory 3D printers utilizing Texas Instruments DLP projectors.

~ ~

ExOne

With so many years of manufacturing experience and considerable outlays in research and product development, ExOne has forged the fruition and advancement of non-conventional manufacturing.

~ ~

Stratasys, Ltd.

A producer of 3D printers and 3D production systems for office-based rapid prototyping and direct digital manufacturing solutions, Stratasys maintains dual headquarters in Eden Prairie, Minnesota and Rehovot, Israel. The company holds nearly 500 granted or pending additive manufacturing patents all over the world. It is a public company that trades on NASDAQ under the symbol SSYS.

Its systems range from reasonably priced desktop 3D printers to huge and highly developed 3D production systems, making 3D printing more accessible than ever. Numerous manufacturers utilize Stratasys 3D printers to generate models and prototypes for innovative product design and testing, and to put together finished goods in low volume. In the field of education, educators use the technology to promote research and learning in science, engineering, design and art; hobbyists and entrepreneurs use Stratasys 3D Printing to spread out manufacturing into the home — creating gifts, novelties, customized devices and inventions while engineers use Stratasys systems to style intricate geometries in a wide variety of thermoplastic materials that incorporate ABS, polyphenylsulfone (PPSF), polycarbonate (PC) and ULTEM 9085.

All Stratasys 3D Printers manufacture parts layer-by-layer. FDM Technology or better known as fused deposition modeling, known for its dependability and sturdy components, extrudes fine lines of molten thermoplastic, which harden as they are deposited. Polyjet Technology, known for its even, detailed surfaces and capacity to combine multiple materials in one part, utilizes an inkjet-style technique to manufacture parts from liquid photopolymers in fine globules instantaneously treated with ultraviolet light. SCP Technology or smooth curvature printing, creates finely detailed models for lost-wax casting and mold-making.

The Stratasys portfolio of specially engineered 3D printing materials is the most wide-ranging in the entire industry. It includes nearly 150 PolyJet photopolymers and FDM thermoplastics. Through its set-up of certified resellers, Stratasys delivers receptive, regional support around the globe.

http://www.stratasys.com/3d-printers

3D Systems

3D Systems Corporation is a holding company incorporated in Delaware and operates through subsidiaries in the United States, Europe and the Asia-Pacific region. It is a leading global provider of 3D content-to-print solutions including 3D printers, print materials, on-demand custom parts services and 3D authoring solutions for professionals and consumers.

The company also offers scanners for an assortment of medical and mechanical X-Ray film digital archiving. Its proficiently integrated solutions restore and harmonize conventional techniques and lessen the time and cost of designing new products. 3D Systems' products are employed to quickly design, communicate, prototype and produce real, functional product components.

3D Systems pioneered 3D printing and digital manufacturing with the discovery and development of stereolithography and the universally utilized ".stl" file format over 25 years ago and afterward developed selective laser sintering, multi-jet modeling and film transfer imaging.

3D Systems is committed to democratizing access and accelerating adoption of its products and services with their very reasonable prices and ease of use for the benefit of all end users. The company extends its range of reasonably priced printing solutions from design department and production floor to classrooms, living rooms and maker spaces.

The company provides its customers with 3D authoring tools for digital imaging and design including 3D CAD modeling, feature capture, manipulation, replication and measurement. Its 3D authoring solutions combine 3D content creation and manipulation via CAD modeling, reverse engineering and inspection software that enable its customers to process 3D scanned data directly within a parametric CAD environment. It also offers proprietary software printer drivers together with pre-sale and post-sale services, ranging from applications development and custom engineered production solutions to installation, warranty and maintenance services.

Latest Developments

In October 2012, 3D Systems obtained Rapidform scan-to-CAD reverse engineering and inspection software tools. In November 2012, it launched its new ProJet 3500 Max professional 3D printers with greater productivity, high definition prints and remote tablet controls for the production of functional plastic parts and investment casting wax patterns for product design and manufacturing applications.

Then in the 4th quarter of 2012, it also launched VisiJet Jewel, VisiJet Pearlstone and VisiJet X print materials. VisiJet Jewel is a specialized material formulated for high volume jewelry production for master models for direct casting of cost effective jewelry on its ProJet 6000 and ProJet 7000 professional 3D printers.

In December 2012, it expanded its Quickparts on demand parts service globally, extending its reach across Europe and Asia-Pacific, and launched its Quickparts proprietary instant online quoting engine in Europe.

In January 2013, it launched two new products -- the second generation home 3D printer, the Cube , and its next generation desktop printer, CubeX. The Cube is a reasonably-priced and simple-to-use 3D printer for both children and adults. The second generation Cube provides silent and faster printing, additional print modes and greater materials selection, including recyclable ABS plastic and compostable PLA plastic in new colors. The CubeX is a desktop 3D printer with the largest print volume in its category, triple color printing and multiple print modes and settings and provides professional and superior-quality printing within a natural consumer experience.

Also in the same month, 3D Systems also launched its new ProJet 3500 series, these are professional 3D printers in eight models which deliver more efficiency and greater productivity in the production of functional plastic parts and investment casting wax patterns for specialized grade design and manufacturing functions. The new ProJet 3510 series integrates its patented Multi-Jet Modeling (MJM) print technology, production-grade print heads which include a 5-year service contract, advanced material management, tablet-like touch-screen controls and remote tablet and smartphone connectivity.

Then in February of the same year, the company announced the availability of Go!MODEL, a 3D reverse engineering and design tool developed in collaboration with portable 3D measurement solutions specialist Creaform. Using the integrated Go!MODEL and Go!SCAN 3D package, users can capture physical objects and directly model high quality renderings and designs that are ideal for 3D printing. Go!MODEL, powered by Rapidform platform, offers easy-to-use and inexpensive, professional mesh editing capabilities with automatic surfacing utility and functionality.

http://www.3dsystems.com/
www.4-traders.com ›
Nyse › 3D Systems Corporation

http://yahoo.brand.edgar-online.com/EFX_dll/EDGARpro.dll?FetchFilingHtmlSection1?SectionID=9110858-9244-61550&SessionID=APoR6q-zAfHZT97

Solidscape

Solidscape Inc. is a major manufacturer of high-precision 3D printers, materials and software for Direct Manufacturing of solid objects designed through the use of CAD.

These wax objects are perfect for lost wax investment casting and mold making applications, offering the highest standards in surface finish, accuracy and material castability, getting rid of the need for post-processing.

Solidscape printers are approved in the production of miniature components and assemblies utilized in consumer electronics, biomedical products, orthopedics, dental prosthetics, orthodontic appliances, jewelry, toys, video games, power generation, and sporting goods.

According to the Wohlers Report, with more than 4,000 systems installed in over 80 countries around the world, Solidscape is the 4th biggest supplier of 3D printers. Sales and support are conducted directly by Solidscape personnel together with an established group of Value Added Resellers.

Solidscape was incorporated in February 1994 with global headquarters in Merrimack, New Hampshire, USA and is a Stratasys, Inc. company since May 2011.

[It should be noted here that Solidscape, Inc. is a subsidiary of Stratasys, Ltd.]

http://www.solid-scape.com/

Arcam

Arcam offers a cost-efficient additive manufacturing solution for production of metal components. Arcam technology puts forward freedom in design merged with outstanding material properties and high productivity. Arcam has an international market with customers principally in the orthopedic implant and aerospace industries.

Founded in 1997 and listed on NASDAQ OMX Stockholm, Sweden, its head office and production facilities are located in Mölndal, Sweden while support offices are situated in the United States, Italy, China and the UK.

Arcam is a pioneering partner for manufacturing in the orthopedic implant and aerospace industries, where it delivers customer value through its competence and solution-orientated operation.

Since its inception, its vision has remained the same and that is to revolutionize the art of manufacturing multifaceted components. Arcam offers a comprehensive portfolio of EBM machines, auxiliary paraphernalia, software, metal powders, service and training to support its customers. Arcam has filed more than 25 patent families for its EBM technology and has currently more than 50 granted patents in several countries.

Http://www.arcam.com

http://www.bonezonepub.com/component/content/article/603-suppliers-and-service-providers-offer-innovative-solutions-

http://www.bonezonepub.com/component/search/?searchword=arcam&ordering=&searchphrase=all

Optomec

Optomec, a lucrative and swiftly growing additive manufacturing firm, spearheads the next generation in electronics, energy, life science and aerospace/defense manufacturing.

A recognized leader in the field of additive manufacturing, the company has invested more than $30 million in the development of solutions that leap-frog contemporary manufacturing capabilities. Major application areas include trimming down the size and cost of electronic devices, enhancing the efficiency for the generation of alternative sources of energy, lengthening the life of high value aerospace constituents, and creating and developing wear-resistant medical devices.

Optomec launched its very first commercial additive manufacturing system in 1997 and has now installed systems at 150 customer sites in 15 countries. Its customer base includes many marquee names in the industry that are utilizing Optomec systems in developing tomorrow's innovative products, which includes leading smart phones, solar and touch screen display manufacturers. Likewise, laboratory and government organizations like National Renewable Energy Laboratory (NREL), Fraunhofer (IFAM, IKTS, ENAS, IWS, ISE), CEA LETI, IMEC, PeTeC, Sirris, NASA, Sandia National Laboratories, as well as the US Air Force, US Army and US Navy, are using its systems to advance a variety of commercial and military applications, such as expanding the functional life of present worn military components.

Optomec is privately held, with headquarters in Albuquerque, New Mexico. The company has field sales and support offices located throughout the United States and a global network of distributors and agents. It sells its products internationally to OEMs, system integrators and end users in an extensive array of diverse markets. Optomec holds substantial Intellectual Property assets, including more than thirty (30) issued patents and another fifty (50) pending, all in the field of additive manufacturing.

www.optomec.com/Company/Overview

http://www.sme.org/uploadedFiles/Publications/ME_Magazine /2013/June_2013/Web-Only_Content/RAPID_8395_Directory_PDF.pdf

Voxeljet

Voxeljet, another leading manufacturer of industrial 3D printing systems, operates what it believes is one of Europe's largest service centers for the on-demand production of molds and models for metal casting.

Voxeljet systems business division focuses on the development, production and distribution of the market's fastest and most powerful 3D printing systems. At present, Voxeljet has a well-coordinated product range that reaches from smaller entry models to large-format machines and therefore offers the perfect 3D print system for many application areas.

Voxeljet's exceptional mishmash of technological know-how and unfailing enhancements in technology makes the company an internationally respected partner for demanding clients from the world of 3D printing.

http://www.voxeljet.de/en/company/

http://www.bullfax.com/?q=node-new-addition-3d-printing-family

EnvisionTEC

With corporate headquarters in Gladbeck, Germany and North American headquarters in Dearborn, MI., EnvisionTEC was founded in 2002 in Marl, Germany. Under the leadership and direction of Mr. Siblani, Chairman of the Board, EnvisionTEC has become a world leader in rapid prototyping and manufacturing equipment.

Focusing in the development, production and sale of cost effective, 3D Printers including software and materials, the company offers branch/customer specific solutions for Solid Freeform Fabrication (SFF) systems. This includes 3D printing applications in toys, sporting goods, medical, manufacturing, jewelry, hearing aids, electronics, education, dental, consumer goods and packaging, automotive, architecture and art, animation entertainment, and the aerospace industry.

Employing a panel of in-house specialists in optical, mechanical, and electrical engineering, EnvisionTEC produces the most reliable 3D Printers in the world.

EnvisionTEC makes use of two types of technology to cure liquid resin into a 3 dimensional object:

DLP

In 2002, EnvisionTEC began creating and generating Perfactory 3D printers utilizing Texas Instruments DLP projectors. Perfactory systems build 3D objects by employing the projectors to project voxel data into liquid resin which then causes the resin to cure from liquid to solid. The ease and straightforwardness of the technology has made the system extremely famous in rapid manufacturing markets like the hearing aid sector where EnvisionTEC enjoys more than 60% share of the world market as well as over 50% of the 3D jewelry printer market in the number of units produced. EnvisionTEC also collaborates with the best in the business to produce comprehensive turn-key solutions for market specific areas. The company effectively combined its technology with 3shape in the Hearing Aid market, and with Dental Wings in the Dental market and many Jewelry design software packages.

3SP Scan, Spin and Selectively Photocure

This is a multi-cavity laser diode with an orthogonal mirror spinning at 20,000 rpm, the light is replicated through the rotating drum and undergoes a succession of optical elements in so doing centers the light onto the surface of the photo polymer across the Y direction.

EnvisionTEC has also integrated the Materialise Magics EnvisionTEC software with its Perfactory Software Suite giving their customers the golden standard in STL file repair and manipulation capabilities. The company has over 90 patents and patent applications awaiting all over the world and it values its intellectual property and continues to issue an average of one patent every three months to protect its inventions in the field of Rapid Prototyping and Manufacturing using selective light modulation which is presented with the DLP technology that it currently employs.

en.wikipedia.org/wiki/EnvisionTEC
envisiontec.com/technology-overview/
http://envisiontec.com/envisiontec/wp-content/uploads/2013/03/VDM-2013.pdf

ExOne

With so many years of manufacturing experience and considerable outlays in research and product development, ExOne has forged the fruition and advancement of non-conventional manufacturing. Such investment has given way to a fresh generation of rapid production technology in the field of additive manufacturing as well as sophisticated micromachining processes.

The company's process solutions give commodity producers the autonomy to manufacture objects that have practically boundless design intricacy. ExOne collaborates with its clients through the full development and production process so that it is able to turn up novel ideas — designs, prototypes, and production parts — specifically when required.

It supports the employment of traditional industrial strength materials ranging from metals to ceramics to glass, all used in revolutionary ways. ExOne's full range of offerings also includes services and equipment for fabricating on a "micro" scale, which facilitates machining of small attributes with accuracy and swiftness. ExOne is the most favorable and most advantageous collaborator for any industrial producer who is shifted their manufacturing business to the digital age.

www.investor.exone.com/releasedetail.cfm?ReleaseID=772032
www.exone.com/

Wrapping Up

At this point in time, the industry is getting so much attention from investors, government agencies and corporations. On the other hand and even with the kind of attention it is getting, some people continue to be skeptical on the potentials of 3D printing and still have reservations on the economic feasibility of the industry. However, this treatise stays on the premise that the 3D printing industry will drastically change the landscape of the manufacturing industry in the next few years.

I

REFERENCES

1. Burnett, Ronald , PhD; (March 2012)
"Manufacturing: 3D printing - a Disruptive technology?"
http://www.economicstraighttalk.com/NewsArticles/Commentary.aspx?Id=1104

2. Geller, E. (Fall 2012),
       "www.file2part.com/IQT_Quarterly_Fall_2012.pdf" Vol. 4 No. 2. 09.

3. Hall, Matthew et all; (July 2013)
"3D Printing - Top Ten IP Challenges"
http://www.acc.com/legalresources/publications/topten/tdpipc.cfm?makepdf=1

4. Hall, Matthew; (May 2013)
"3D Printing - Top 10 IP Challenges"
       http://www.swaab.com.au/knowledge/publications/3d-printing-top-10-ip-challenges/

5. Savitz, Eric; (December 2012)
"Manufacturing The Future: 10 Trends To Come In 3D Printing"
http://www.forbes.com/sites/ciocentral/2012/12/07/manufacturing-the-future-10-trends-to-come-in-3d-printing/

6. Barcelona, Moebio; (April 2013)
"Printing a medical revolution"

http://blog.moebio.org/2013/04/25/printing-a-medical-revolution/

7. Cooke, A.L., Moynihan, S.P.; (August 2011)
"PROCESS INTERMITTENT MEASUREMENT FOR POWDER-BED BASED
ADDITIVE MANUFACTURING"
http://utwired.engr.utexas.edu/lff/symposium/proceedingsArchive/pubs/Manuscripts/2011/2011-07-Cooke.pdf

8. Wikipedia, "Selective Laser Sintering"
http://en.wikipedia.org/wiki/Selective_laser_sintering

9. Wikipedia, "3D Printing"
http://en.wikipedia.org/wiki/3D_Printing